José Luis Rios Flores
Sigifredo Armendáriz Erives
Manuel de Jesús A. Ruiz-Esparza

Water footprint, using water productivity indices

José Luis Rios Flores
Sigifredo Armendáriz Erives
Manuel de Jesús A. Ruiz-Esparza

Water footprint, using water productivity indices

In wheat(Triticum aestivum L.) versus chickpea (Ciser arietinum L.) in Cajeme, Sonora

ScienciaScripts

Imprint

Any brand names and product names mentioned in this book are subject to trademark, brand or patent protection and are trademarks or registered trademarks of their respective holders. The use of brand names, product names, common names, trade names, product descriptions etc. even without a particular marking in this work is in no way to be construed to mean that such names may be regarded as unrestricted in respect of trademark and brand protection legislation and could thus be used by anyone.

Cover image: www.ingimage.com

This book is a translation from the original published under ISBN 978-620-2-14522-0.

Publisher:
Sciencia Scripts
is a trademark of
Dodo Books Indian Ocean Ltd. and OmniScriptum S.R.L publishing group

120 High Road, East Finchley, London, N2 9ED, United Kingdom
Str. Armeneasca 28/1, office 1, Chisinau MD-2012, Republic of Moldova, Europe
Printed at: see last page
ISBN: 978-620-7-67247-9

Contents

SUMMARY

The objective was to determine the water footprint of wheat grain and white chickpea crops produced in Cajeme Sonora, for which mathematical models were developed to determine the efficiency and productivity of water in both crops. It was determined that a total of 1,056 L kg^{-1} was used for wheat and 1,621 L kg^{-1} for white chickpea. The physical productivity indicators were 0.947 kg m-3 in wheat per cubic metre and 0.617 kg m-3 in white chickpea. In economic terms, a total of 115.63 m^{3} of water was used in wheat to generate US\$ 1 dollar of loss, and a total of 6.84 m^{3} per dollar of profit was used in chickpea. The jobs generated per cubic hectometre of water used in irrigation were 0.56 jobs hm^{3} in wheat and 1.16 jobs hm^{3} in white chickpea. Likewise, 1.4 h ton^{-1} in wheat grain and 4.3 h ton^{-1} in white chickpea were required. It is concluded that the white chickpea produced in Cajeme, Sonora showed higher economic and social efficiency and productivity indicators than those determined for the wheat crop.

Keywords: Virtual water, water productivity, water efficiency, water footprint.

CHAPTER I

I. INTRODUCTION

Irrigated agriculture in the country has been established mostly in arid and semi-arid zones, and for this reason a set of hydraulic works have been built to store, light and distribute the water required by agricultural crops during their growth. Currently, the challenge facing this activity is to achieve higher efficiency rates of the derived volumes, in order to strategically combat the growing danger of having less availability in quantity and quality of water for different uses (CONAGUA, 2007)[1] .

Agriculture today faces several challenges of economic and ecological sustainability. In this context, irrigation areas in Northwest Mexico, especially pumped irrigation, need to make a more efficient use of resources, mainly water, as well as to increase their productivity and efficiency (INIFAP, 2004)[1,2] . In this sense, the northwestern regions of Mexico, particularly the state of Sonora, with a great cereal tradition, where 53% of its agricultural surface is planted with wheat and chickpeas. Particularly in the District of Cajeme, Sonora, SIAP (2014) reported a total of 269,323.69 hectares, of which 73.70% were planted with wheat and chickpeas, representing 53.72% of the Gross Value of Production in this agricultural region.

However, according to INIFAP (2004)[3] , at the state level, producers have had to balance their agricultural activity with other crop options, such as fruit and vegetables that can be traded on international markets. In this sense, chickpea has agronomic characteristics that give it advantages over other crops such as wheat, which is why it has become one of the economic pillars of the northwest area, since it combines low water requirements with a good adaptation to the desert climate, in addition to having good market prices, It has therefore been considered a good option for production in this region (INIFAP, 2004)[4] , and FAO (2003)[5] , mentions that in order to achieve a better

[1] CNA. 2007. National Water Commission. Determination de la Disponibilidad de Agua en el Acui'lero 2605 Caborca, Estado de Sonora. CONAGUA, 31p.
[2] INIFAP. 2004. The cultivation of white chickpea in Sonora. Instituto Nacional de Investigaciones Forestales, Agricolas y Pecuarias. Northwest Regional Research Centre. Costa de Hermosillo Experimental Field. Libro Tecnico No 6. 290p.
[3] INIFAP. 2004. Op cit.
[4] INIFAP. 2004. ^dem.

economic and social use of water, methods are needed to evaluate its productivity, in order to make better decisions regarding sustainable use policies and strategies.

[5] FAO: Unlocking water's potential for agriculture Chapter 3. Why water productivity matters for the global water challenge. Ed. Sustainable Development Department. Rome, Italy. 2003.

II. OBJECTIVE AND HYPOTHESIS

2.1 Objective

The *general* objective was to determine the ttdric footprints, through numerical indicators of efficiency (amount of irrigated water per kg of wheat and chickpea produced) and productivity (kg of wheat and chickpea produced per m^3 of irrigated water) of surface water irrigated by gravity in the cultivation of wheat grain and white chickpea in the region of Cajeme, Sonora and once they have the *main specific* objective of comparing them with each other to determine the degree of physical, economic and social efficiency in water use in each crop is raised, The *secondary* specific objectives are to generate indicators of the social productivity of capital and labour productivity in each of the two crops, in order to compare them with each other and thus determine which of the two crops is more productive in the use of capital and labour force.

2.2 Hypotheses

First hypothesis: The water footprint in **physical** terms, measured as L kg^{-1} , and kg m^{-3} of wheat grain is *higher* than the corresponding water footprint of white chickpea.

Second hypothesis: The water footprint in **economic** terms, measured as m^3 /US\$ profit, US\$ hm^{-3} of the white chickpea crop is *higher* than that of the wheat grain crop.

Third hypothesis: In the DDR Cajeme, the **social** footprint, measured as jobs hm^{-3} of the white chickpea crop is higher than that of the grain wheat crop.

III. LITERATURE REVIEW

3.1 Irrigation District 041 and Water Sustainability

Irrigation District 041, Rfo Yaqui, in northwestern Mexico (Fig. 1a and 1b), is a region that in recent years has been affected by unsustainable agricultural development. The establishment of intensive agriculture, coupled with a prolonged drought, collapsed the dam system and, consequently, the agricultural activity of the irrigation district in the 2002-2003 agricultural year. During this period of drought, both the Yaqui River dam system and the aquifers of the Yaqui and Cocoraque Valleys proved to be extremely vulnerable, and it is necessary to take measures to increase the efficiency in the management and operation of water resources and, in this way, to achieve sustainable use of water and the activities that depend on it.

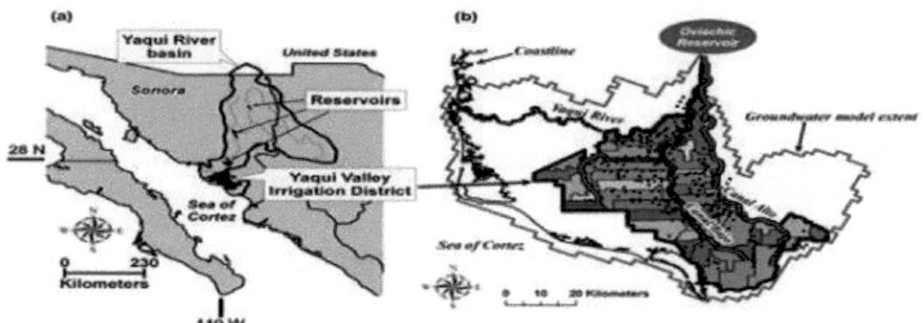

Fig. 1. Localization of the Yaqui Valley: Yaqui River basin (a) and (b) Irrigation District 041. Source Schoups *et al.*, (2006)6. [6]

[6] Schoups, G., Addams, C. L., Minjares, J. L., & Gorelick, S. M. (2006). Sustainable conjunctive water management in irrigated agriculture: Model formulation and application to the Yaqui Valley, Mexico. Water Resources Research, 42(10).

In general, agriculture in the northwest is almost entirely irrigated, which leads to the use of both surface and groundwater, but one of the main challenges is to improve irrigation efficiency (Salinas-Zavala *et al.*, 2006)[7]. The dominant crop is winter wheat, which is grown from November to April and is irrigated using a combination of surface and groundwater. The surface water system consists of three reservoirs in the Yaqui with a total capacity of 7000 m3. Average annual runoff from the No Yaqui is 2, 700 106 m3, generated by lower evaporation rates and higher precipitation rates at higher elevations in the No Yaqui basin.

The severe restrictions on the volumes of water authorised for pumping and the farmers' search for crops that generate more profit have made it necessary to convert to crops that consume less water and generate a high profit. In this sense, the chickpea crop has both characteristics, on the one hand its low water requirements in the northwest region, which range between 35 and 70 cm of irrigation lamina (INIFAP, 2004)[8] [9] . In comparison with crops such as wheat in Cajeme, Sonora, in gravity irrigation uses 97cm of irrigation sheet, while in Caborca, in pump irrigation requires 96cm, on the other hand in the region of Baja California Sur the irrigation sheet of this crop in gravity is 97cm, so that according to the calculations at regional level this crop used in the 2012 agricultural cycle a total of 2,433.71 Mm^3 (Amaya, 2015)[9] . Another crop that uses a large amount of water for irrigation in the northwest is cotton, which uses an irrigation sheet of just over 100cm, according to Rtes *et al.*, (2014)[10] , used a total of 126.60Mm3 in Cajeme Sonora, with an irrigation sheet of 1.07m. Grape cultivation in the northwest is also one of the crops that uses a large amount of water, according to Rtes *et al.*, (2014a)[10] [11] , with which it is

[8] Salinas-Zavala, C. A., Salvador E, L. C., & Fogel, I. 2006. History of winter wheat crop development in five irrigation districts in the Sonoran Desert, Mexico. Interscience. 31(4): 254-261.

[9] INIFAP. 2004. White chickpea cultivation in Sonora. Instituto Nacional de Investigaciones Forestales, Agricolas y Pecuarias. Northwest Regional Research Centre. Costa de Hermosillo Experimental Field. Libro Tecnico No 6. 290p

[9] Amaya, M. J. 2015. Grain wheat (Triticum vulgare) from Cajeme, Sonora and its hydraulic footprint. Thesis. Universidad Autonoma Chapingo- Department of Zootechnics. Texcoco Edo de Mexico. 55p.

[10] Rios, F. JL.; Torres, M. M.; Ruiz, T. J.; Castro, F. R. 2014. Productivity and irrigation water efficiency in cotton (Gosspium hirsuntum) from DR-017 Comarca Lagunera and DDR-148 Cajeme, Sonora. Report of the X Congreso Nacional Sobre Recursos Bioticos De Zonas Aridas. Bermejillo, Durango 1 October 2014. 206-214pp.

[11] Raos F. JL.; Torres, M. MA.; Torres, M. M.; Castro, F. R. 2014a. Economic and social productivity of water in sultana and table grapes in DDR-037, Sonora. Report of the 59th Annual Meeting of the Central American Cooperative Programme for Crop and Animal Improvement (PCCMCA). Managua, Nicaragua from 28 April to 3 May 2014. 128pp.

estimated that during the 2012 agricultural cycle these crops used a total of 81.28 Mm^3 in the Irrigation District 037-Altar- Pitiquito-Caborca, Sonora. Despite the fact that water is extremely scarce in these regions, water prices are extremely low, $0.17 m-3 cotton (Rtes *et al.,* 2014)[12] and $0.16 m^{-3} for wheat in Cajeme, $0.71 m^{-3} for wheat in Caborca, and $0.19 m^{-3} for wheat produced in Baja California (Amaya, 2015)[13] . While in table grapes and sultanas for Irrigation District 037 the price per cubic metre was equal to $0.95 m .$^{-3}$

The methodology for the determination of the water footprint is recent, it was developed in the United Kingdom and at the University of Netherlands, in Holland, its main representatives are Kijne, Barker and Molden (2003)[14] , Hoekstra (2006)[15] , Hoekstra and Chapagain (2007)[16] , Hoekstra, Chapagain, Aldana and Mekonnen (2011)[12 13 14] , Mekonnen and Hoekstra (2010)[15] , In the Mexican case, it is the IMTA (Mexican Institute of Water Technology) who applies it at the watershed level, and although it has hardly been applied in Mexico for its practical application in the generation of water-saving alternative agricultural patterns that maximise production value, profits and employment, some authors such as Rios *et al* (2015)[16] have already applied it in the case of DR-017 of La Comarca Lagunera, determining firstly the water footprint of the crops of the current agricultural pattern, then, secondly selecting the best crops, i.e. those with the lowest physical, economic and social water footprint, generated alternative cropping patterns that reduced by up to 492 Hm^3 (one

[12] Rios, F. JL.; Torres, M. M.; Ruiz, T. J.; Castro, F. R. 2014. Productivity and irrigation water efficiency in cotton (Gosspium hirsuntum) from DR-017 Comarca Lagunera and DDR-148 Cajeme, Sonora. Report of the X Congreso Nacional Sobre Recursos Bioticos De Zonas Aridas. Bermejillo, Durango 1 October 2014. 206-214pp.

[13] Amaya, M. J. 2015. Grain wheat (Triticum vulgare) from Cajeme, Sonora and its hydraulic footprint. Thesis. Universidad Autonoma Chapingo- Department of Zootechnics. Texcoco Edo de Mexico. 55p.

[14] Kijne, J. W; R. Barker; Molden, D (eds.) 2003. Water productivity in agriculture: Limits and Opportunities for Improvement. CABI Publication, Wallingford UK. 322p.

[12] Hoekstra, A. Y. 2006. The global dimension of water governance: Nine reasons for global arrangements in order to cope with local water problems. Value of Water Report Series No. 20. UNESCO-IHE, Delft, Netherlands. Available at: http://www.waterfootprint.org/Report_20Global_Water_Governance.pdf/

[13] Hoekstra, A. Y.; Chapagain, A. K.; Waterfootprints of nations: Water use by people as a function of their consumption pattern. Water Resources Management. 21(1):35-48.

[14] Hoekstra, A. Y.; Chapagain, A. K.; Aldaya, M.; Mekonnen, M. M. 2011. The water footprint assessment manual setting the global standard. Earthscan, London, UK. 227p.

[15] Mekonnen, M. M.; and Hoekstra, A. Y. 2010. A global and high-resolution assessment of the Green, blue and grey water footprint of wheat. Hydrology and Earth System Sciences. 14: 1259-1276.

[16] Rios, J. L.; Torres, M. M.; Castro, R.; Torres, M. A. 2015. Determination of the blue water footprint in forage crops of DR-017 Comarca Lagunera, Mexico. In: Revista de la Facultad de Ciencias Agrarias. National University of Cuyo. Volume 47. No. 1 year 2015. ISSN printed 0370-4661. ISSN on-line 1853-8665. Mendoza, Argentina. pp. 93-108.

Hm^3 equals one million m $)^{317}$ the volume of water used in agricultural production, while managing to determine that the GVP (Gross Value of Production), profits and employment in the Irrigation District not only did not decrease, but increased by up to 30.6%, 90.4% and 13% respectively.

3.2 Wheat and chickpea production in Sonora

Until before 1999, groundwater availability was approximately 1,300 Mm^3 , while extraction was 1,800 Mm^3 , causing an overexploitation of the aquifers (Donnadieu *et al*, 1999)[18][19] . In later studies for the aquifers of Sonora, it was determined that the situation was more critical, as for the Yaqui Valley an extraction of 560 Mm^3 and a recharge of 460 Mm^3 was estimated, which led to strong restrictions in the volumes authorised for pumping and the search for crops that were more profitable, This is why the chickpea crop met the required characteristics, since on the one hand, its low irrigation requirements were demonstrated in comparison with wheat, and on the other, the great demand for chickpeas in the foreign market (Monreal, 2002)[22].

Since the beginning as a green revolution site in the 1950s and 1960s, wheat cultivation has dominated the cropping system thanks to favourable biophysical conditions for its cultivation, and with the availability of irrigation water. Cotton, soybean and corn crops played important roles during the spring-summer growing season, which has been largely vacant since 2002 due to drought. In recent years, declining wheat prices, rising production costs and concerns about freshwater availability have threatened agriculture, so state and federal authorities, along with progressive farmers, began to explore and promote crop diversification as a strategy to stimulate economic growth in the Yaqui Valley and promote water productivity (McCullough and Matson, 2011)[23].[20]

---

[17] It should be mentioned that according to the authors, they point out in their work that in the DR-017 Comarca Lagunera, an annual extraction of groundwater is estimated at around 1.1 Hm^3 , while the annual recharge of the aquifer is 0.45 Hm^3 , thus bringing extraction and annual recharge closer together, which suggests an alternative for sustainability in regional agricultural production.

[18] **Donnadieu P**, M Roux-Michollet, V Chastagnier . 1999. Philosophical Magazine A, **1999^Taylor** & Francis. Available in:

A quantitative study by **transmission electron** microscopy of nanoscale precipitates in Al-Mg-Si alloys

[19] Monreal, R., M. Rangel, J. Castillo and M. Morales. 2002. Estudio de cuantificacion de la cuanficacion de la recarga del acuifero de la Costa de Hermosillo, municipio de Hermosillo, Sonora, Mexico. Hermosillo: University of Sonora.

[23] [20] McCullough, E. B., & Matson, P. A. (2011). Evolution of the knowledge system for agricultural development in the Yaqui Valley, Sonora, Mexico. Proceedings of the National Academy of Sciences, 201011602. 1-6p.

In the face of NAFTA, there are three possible survival strategies for national producers, especially traditional exporters such as those in Sonora: lowering production costs, increasing yields and diversifying crop patterns. The technology of furrow wheat and conservation tillage as a low-cost, labour- and cost-saving option has spread in the Yaqui Valley; even some producers are reluctant to adopt it. Costs have been reduced on average by up to 40 percent, but the adoption of this technology is a function of cost reduction, not of the conviction of its productive value, which is also proven to yield the same as conventional technologies. Diversification with vegetables, corn and cotton has been proposed, but there is still no crop that can replace the 200,000 hectares planted with wheat in southern Sonora, and it is grown despite the differences in production costs with the main competitors. For its part, the history of vegetables has not changed substantially since before NAFTA was signed. Among the most important vegetables are: watermelon, asparagus, lettuce, cauliflower, chilli, tomato, melon, onion and courgette. But there are researchers who speak of more than 100 potential crops for the area. The largest increase in cultivated area of vegetables was recorded between 1975 and 1990; the percentage of total cultivated area was 1.46% and 5.33%, respectively. Exports rose in the same periods from 4.5% to 14% of the total; Sonora ranked second in horticultural exports (Sanchez, 2008)24.[21]

Wheat in Sonora is not only a crop and an economic activity of the rural producers, its production is immersed in the culture itself; the Yaqui Valley is a world reference on the production of the cereal. The climatic conditions of the state allow 75% of wheat production to be of the crystalline type, as flour wheat has lower productivity and is more susceptible to diseases. Wheat in Sonora represents 38% of the cultivated area in Mexico and 50% of the national production. Yields are 30% above the national average and 121% above world yields, according to data from the Food and Agriculture Organisation of the United Nations (FAO), where some producers in the Yaqui Valley have recorded yields of up to 11 tonnes per hectare. These results are feasible because the state has the best climatic conditions for wheat

---

[24] [21] Sanchez, M. A. (2008). Social sector organisations in the Yaqui Valley. Frontera Norte, 20(40): 135-167.

production, with favourable circumstances in autumn and winter to provide at least 750 hours of cold. Sonora's wheat has three well-defined markets: one is the flour industry, which consumes 23% of the state's production; the other is the livestock market, mainly the pig industry, which demands 31%; and lastly, exports, with 46%. Of the three, the flour industry represents a clear opportunity for supplier development, linking small producers with the market (Diaz, 2013)[25].[22]

In terms of harvested area, chickpea represents the sixth most important crop in the state, after grain wheat (50.3%), safflower (5.8%), grain sorghum (5.3%), alfalfa (4.5%), grain maize (3.7%) and grain chickpea (3.0%), also representing 1.1% of the Gross Value of the state production generated in the 2014 agricultural cycle. The importance of the crop in recent years is due to its profitability and efficiency in water use (35-70 cm of irrigation sheet), as well as the generation of labour and foreign exchange (INIFAP, 2005)[23] . In the state of Sonora, the chickpea crop is mainly produced under irrigation (99%), while the remaining 1% is produced under rainfed conditions, and 99.6% of the production is obtained from irrigation, which indicates that 99.7% of the Gross Value of Production is generated under irrigation. Planting dates extend from 15 November to 15 January, with yields ranging from 1.34 ton ha^{-1} in irrigation to 0.7 ton ha^{-1} in rainfed (SIAP, 2014)[24] .

The Cajeme Rural Development District has been the most productive in history, and in 1990, 5,562 hectares were devoted to vegetable production. The area exploited by horticulturists at the end of 1990 in Sonora reported a value of 47 percent of the total foreign exchange of the agricultural sector, although wheat occupied 40 percent with 224 thousand tons, of which almost 45% corresponded to the Yaqui Valley, since between 1997-1998 the area increased to 10, 446, which represented approximately 60%. The horticultural area represents only 3% of the total and remained the same during the 1990s

---

[25] [22] Diaz, J.P. 2013. Mexican wheat comes from southern Sonora. Economist article of 14 August 2013. Available at: http://eleconomista.com.mx/columnas/agro-negocios/2013/08/14/trigo-mexicano-viene-sur-sonora / last accessed 15 October 2015.

[23]INIFAP. 2005. Coast 2005. New white chickpea variety for the Hermosillo Coast. Instituto Nacional de Investigaciones Forestales, Agricolas y Pecuarias. Northwest Regional Research Center. Costa de Hermosillo Experimental Field. Technical Leaflet No 28. 24p. [27]SIAP, 2014. Anuarios estadisticos de la produccion agropecuaria. Agrifood and Fisheries Information Service. SAGARPA-SIAP. Available at: http://www.siap.gob.mx/ / (last accessed August 25, 2015).

(Sanchez, 2008)28.[25]

3.3 Agricultural production in Cajeme, Sonora

To understand the importance of wheat and chickpea cultivation in Cajeme, one must first understand the importance of Cajeme within the state of Sonora. The state of Sonora is divided according to SIAP (2015) into twelve irrigation districts (Agua Prieta, Caborca, Cajeme, Guaymas, Hermosillo, Magdalena, Mazatan, Moctezuma, Navojoa, San Luis Rfc "s Colorada, and Ures), according to this same source during the 2014 agricultural cycle a total of 526, 662.36 hectares at state level, of which 523, 750.62 hectares were harvested, with a Gross Value of Production (GVP) of $16,920,026.39 thousand pesos. In relative terms, Cajeme represented 51.1% of the agricultural area (269,323.69 hectares), 51.4% of the harvested area and 43.5% of the GVP ($7,354,578.58 thousand pesos), which indicates the importance of the district of Cajeme in the state agricultural production.

According to SIAP (2014), the municipality of Cajeme represented 17.3% of the agricultural surface at state level with 102, 835 hectares of crops, occupying the third place in economic importance by generating 3,450 million pesos (MDP), preceded by Hermosillo and Caborca. They also mention that the main crops of economic importance are wheat grain which in 2014 represented 27.2% of the value of the state production with $7,384 MDP with 2,089,981 tons produced, followed by grapes, which represented 19.5% of the value of production with $5,282 MDP with its 271,580 tons produced. Asparagus, which represents 11.5% of the gross value of production at the state level by generating $3,114 MDP with the sale of 84,023 tons of asparagus. Another economically important crop for the state of Sonora is potato with 8.9%, generating $2,415 MDP with the 352,050 tons generated. Chickpea, which generated $863MDP representing 3.2% of the gross value of production for the state of Sonora (Fig. 2).

---

[25]28 Sanchez, M. A. (2008). Social sector organisations in the Yaqui Valley. Frontera Norte, 20(40): 135-167.

Fig. 2. Leading municipalities and products of the agricultural sector in the state of Sonora. Source: SIAP (2014)[29] .[26]

According to figures from SIAP (2015)[27] , for the period between 2003 and 2014, the area of grain chickpea has experienced oscillations in the harvested area from 218 hectares in 2003 to 1, 817 hectares in 2014, increasing the area at a rate of 19.3% per year, with an average of 2, 555 hectares harvested. Production grew at a rate of TAC=18.9% from 327 to 2, 618.80 tons, while the average rural price went from $3,850 to $11, 287 per ton, which indicates a growth in the order of TAC= 9.4% in the period. Yields for this region averaged 1.83 tons per hectare, while the national average is 1.61 tons per hectare, which indicates that the yield of the Cajeme region is 14% higher than the national average (Fig. 3).

---

[26] SIAP (2014). Food Infographics. Sonora. Agricultural Information System -SAGARPA. 1ª edition. 70pp. Available at: http://www.siap.gob.mx/pdfjs/web/viewer.php?file=Sonora.pdf / (Accessed 20 September 2015).

[27] SIAP. (2015). Agricultural Information System. Cierre de la producción agricola por estado. Available at: http://www.siap.gob.mx/cierre-de-la-produccion-agricola-por-estado/ (Accessed 30 September 2015).

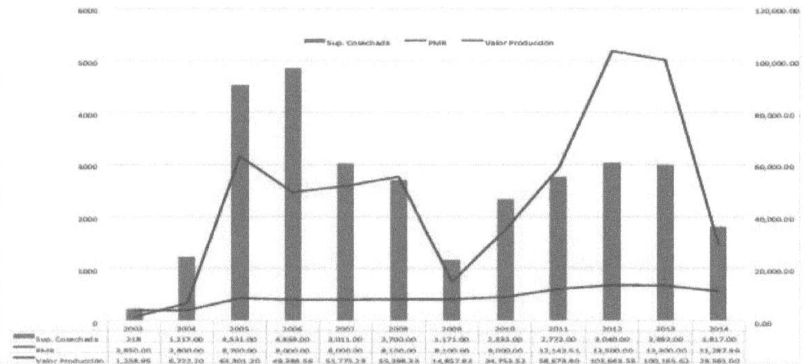

Harvested area, average rural price and gross value of

chickpea production in Cajeme, Sonora.

According to figures from SIAP (2015)[28] , for the period between 2003 and 2014, the area of wheat grain went from 78, 681 hectares in 2003 to 76, 183 hectares in 2014, decreasing the area at a rate of 0.3% per year, with an average of 59,618 hectares harvested. Production grew at a rate of TAC=1.6% from 393,005 to 473,694.90 tonnes, averaging 371,418 tonnes per year, while the average rural price increased from $1,400 to $3,225 per tonne of wheat, which indicates a growth of around TAC=7.2% in the period. Yields for this region averaged 1.8 tonnes per hectare, similar to the chickpea yield, while the national average yield is 5.19 tonnes per hectare, indicating that the Cajeme region yields 84% higher than the national yield (Fig. 4).

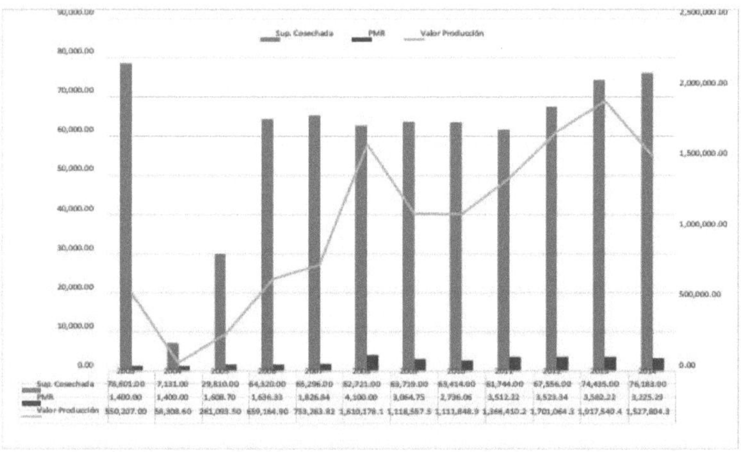

Fig. 4. Harvested area, average rural price and gross value of wheat production in

Cajeme, Sonora.

[28] SIAP. (2015). Agricultural Information System. Cierre de la producción agricola por estado. Available at: http://www.siap.gob.mx/cierre-de-la-produccion-agricola-por-estado/ (Accessed 30 September 2015).

IV. MATERIALS AND METHODS

4.1 Sources of information

The base data used to obtain each and every one of the other variables were the figures for harvested area, annual physical production, prices per ton (dividing the Gross Value of Production by annual physical production), reported for the DRR- 148 Irrigation District, Cajeme Sonora, by SIAP for the 2014-15 agricultural cycle, from SIAP, and the costs per hectare and number of day labourers per hectare reported by FIRA in 2014, obtained from SIAP, Cajeme Sonora, by SIAP for the 2014-15 agricultural cycle, from SIAP and the costs per hectare and number of day labourers per hectare reported by FIRA in 2014, obtained through the Agricultural Cost Elaboration System in its Agricultural Module of FIRA for the cultivation of wheat grain and white chickpea.

The data sources are secondary, since data from the Statistical Yearbook of Agricultural Production, cycle 2014-15, of SIAP (Agricultural and Fisheries Information System) were used, specifically figures on harvested area, annual physical production, average rural price (PMR), and physical yields per ha for wheat and chickpea grain, Average Rural Price (PMR), and physical yields per ha for the wheat and chickpea grain crops. The second source is FIRA, which obtained the total cost per hectare, with each of its constituent items, through its web page: soil preparation, sowing and fertilisation, cultivation work, irrigation, plant health, harvesting, financial and miscellaneous costs. Likewise, the production costs per hectare of FIRA, considers the net volume of water irrigated to the crop, so to obtain the net irrigation sheet, that volume of water was divided by 10,000, so the net sheet was given in linear meters, and as FIRA manages the volume already net applied per hectare to the crop, the irrigation sheet was not subjected to any percentage of efficiency, it is insisted, as FIRA consigns it already as net volume of water applied.

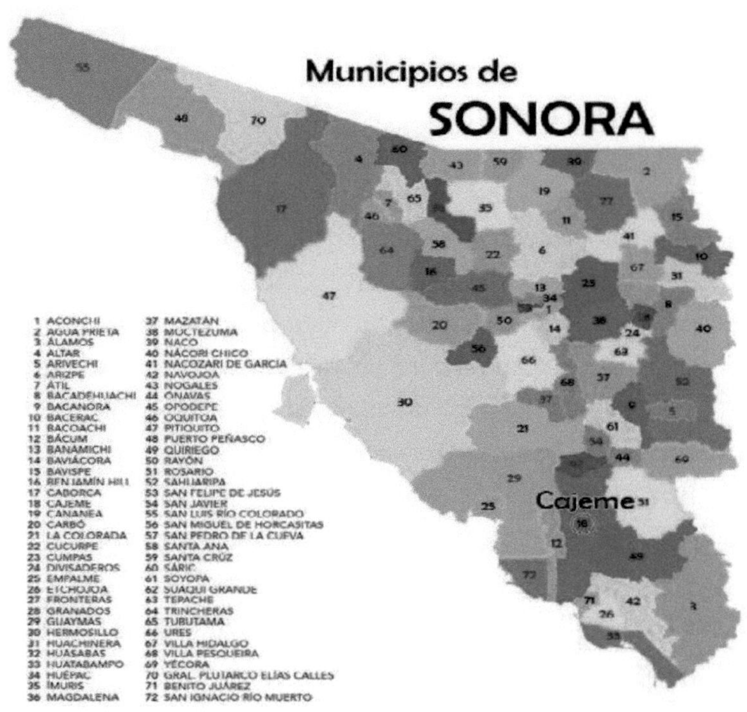

1	ACONCHI	37	M AZATAN
2	EARLY WATER	38	MOCTEZUMA
1	ALAMOS	39	NACO
4	ALTAR	40	NACORI CHICO
2	ARFVECHI	41	NACOZARI DE GARCIA
6	ARtZPE	42	NAVOJOA
7	ATIL	43	WALNUT TREE
8	BACADEHUACHI	44	ONAVAS
9	BACAN ORA	45	OPODEPE
10	BACERAC	46	OQUITOA
11	BACOACHI	47	PITIQUITO
12	BACUM	48	PORT PENASCO
13	BANAMICHI	49	QUIRIEGO
14	BAVIACORA	50	RAYON
15	BAVISPE	51	ROSARIO
16	BENJAMIN HILL	52	SAHUARIPA
'7	CABORCA	53	SAINT PHILIPPE OF JESUS
If	CAJEME	54	SAN JAVIER
• 9	CANANEA	55	SAN LUIS RIO COLORADO
30	CARBO	56	SAN MIGUEL DE HORCASITAS
21	LA COLORADA	57	SAN PEDRO DE LA CUEVA
22	CUCURPE	58	SAN IA ANA
23	CUMPAS	59	SANTA CRUZ
24	DIVISADEROS	60	SARIC
25	EMPALME	61	SOYOPA
26	ETCHOJOA	62	SUAQUI BIG
27	FRONT ERAS	63	TEPACHE
28	GRANADOS	64	TRENCHES
29	GUAYMAS	65	TUBUTAMA
30	HERMOSILLO	66	URES
31	HUACHINERA	67	VILLA HIDALGO

1	ACONCHI	37	M AZATAN
38	HUASABAS	68	VILLA PESQUEIRA
33	HUATABAMPO	69	YECORA
34	HUE PAC	70	GRAL PLUTARCO ELI AS CALLES
35	IMURIS	71	BENITO JUAREZ
36	MAGDALENA	72	SAN IGNACIO RIO MUERTO

Figure 5: Geographical location of the municipality of Cajeme in the state of Cajeme.

Sonora, Mexico (Source: https://dinosenglish.edu.vn/mapa-de-sonora-con-division-politica-1690605724736795/)

The methodology of Rios *et al* (2015)32[39] is used for the estimation of the water footprint through indicators of physical, economic and social productivity of water used in production, which has the advantage of being applied at various levels of aggregation, in which the crop, the geo-economic region, the type of land tenure, the type of irrigation, and even in the same type of irrigation vary, the pumping of underground water for example, the impact of each of these different types of irrigation such as pumping and subsequent gravity irrigation, or the cintilla method, or sprinkler irrigation, canon, central pivot, as each of these types of irrigation in particular, and even more the variation in the type of land tenure or the agricultural region affect differently in the hydraulic footprint. The water footprint methodology was used in this work, in principle, because it is of utmost importance, for the sake of long term sustainability, to know how much water is demanded in the production of any product or service, and in particular, because it is necessary for decision makers such as CONAGUA, SAGARPA, among others, to have indicators that allow the reallocation of scarce resources such as water and soil, always having in mind the minimisation of water use while optimising VW, SAGARPA, among others, to have indicators that allow for the reallocation of scarce resources such as water and soil, always with a view to minimising water use while optimising GVP, profits and employment, i.e. sustainability in the use of available resources.

---

[39]32 Rios, J. L.; Torres, M. M.; Castro, R.; Torres, M. A. 2015. Determination of the blue water footprint in forage crops of DR-017 Comarca Lagunera, Mexico. Journal of the Faculty of Agricultural Sciences. National University of Cuyo. Volume 47. No. 1 year 2015. ISSN printed 0370-4661. ISSN on-line 1853-8665. Mendoza, Argentina. pp. 93-107.

4.2 Delimitations, variables assessed, definitions used in this study and level of aggregation of the analysis.

Based on Mekonen and Hoekstra (2011)[40] , who classify the water footprint into blue, green and grey, the first classification corresponds to the water footprint related to surface water or groundwater or gravity water, as Mekonnen and Hoekstra (2011, *op. cit.)* call it. *Cit*) to groundwater or gravity water, while the grey water footprint is related to wastewater from industry or the urban sector used in agriculture, so in our case it is the blue water footprint, which can be estimated by numerical indicators of water productivity and/or efficiency, where both indicators, has in its primary origin the operation of a quotient, in which, in its form of *efficiency* indicator, records in the numerator the amount of water used (litres, m^3 , gallons) while the denominator records the unit of physical product (kg, pounds, bushels, ton, etc.) produced, while the water footprint, seen as a *productivity* index*, the* numerator and denominator are inverted, with the quantity of physical product at the top of the quotient and the quantity of water used in the denominator; however, in this study we intend to take a step forward and evaluate the water footprint using economic indices, in which the product includes its economic facet, thus, indicators of profit $/m^3$ will be determined, as well as indicators of the water footprint in its social aspect, through the generation of indicators of number of jobs/hectometre, thus having this work the contribution of indicators that serve for the evaluation of the economic and social sustainability.

Based on Hoekstra and Chapagain (2007 *op cit),* the water footprint is *generally* defined as the amount of freshwater used to produce a unit of product or service, for example, every time we drink a cup of coffee is equivalent to consuming 140 litres of water, since that is precisely its *water footprint*, i.e. the average (globally, regionally and temporally) of water required to produce the coffee beans needed to produce a cup of coffee is 140 litres of water, at regional and temporal level varies) that water is required to produce the coffee beans needed to produce a cup of coffee are 140 litres of water, or eating a chocolate bar is the same as consuming 2,400 litres of water,

---

[40] Mekonnen M. M. and A. Y. Hoekstra. 2011. The green, blue and grey water footprint of crops and derived crop products. Hydrology and Earth Systems Sciences 15:1577-1600.

because that is its global average water footprint, eating a big red tomato requires 180 litres of water (we insist, global average, because the water footprint varies from region to region, and even in the same region depending on the type of soil, the production system, the weather...)...), then, in general, the water footprint is nothing more than the amount of water used in the production of a product, and if that water comes from rainfall (i.e. from agricultural production under rainfed conditions) then we speak of the *green water footprint^* while the *blue water footprint* refers to the use of water either underground or of surface origin that was previously stored in dams and distributed to the crops, while the *grey water footprint* refers to the volume of water polluted in the production process and returned to the environment.

To achieve the objective, fourteen independent **variables** were evaluated, reflecting the water footprint in its physical (Y1 and Y2), economic (Y3 to Y7) and social (Y8 to Y14) aspects:

Y_1 = Litres of water/kg

Y_2 = Kg/m^3 of water

Y_3 = Litres of water per \$1 of gross income

Y_4 = Gross revenue per m^3 of water

Y_5 = Gross profit per m^3 of water

Y_6 = Litres of water per \$1 of profit

Y_7 = Price per m^3 of water to the producer

Y_8 = Gross profit per m^3 /price of water m^3 to producer

Y_9 = Jobs generated per 100,000 m^3 of water

Y_{10} = Working hours invested per tonne

Y_{11} = Earnings per worker (\$ thousands)

Y_{12} = Earnings per hour of work invested

Y_{13} = Break-even point (ton/ha to neither lose nor gain)

Y_{14} = Credit vulnerability= Physical yield per $_{ha/Yi3}$ = vulnerable if Y14<1

The study was *limited*, in principle, to the cultivation of **wheat grain**, and **geographically** the study was limited to the Rural Development District (RDD) of Cajeme, Sonora, also limited to the analysis of the type of irrigation by **pumping**, so the study was limited to the determination of the **blue** water footprint, and the crop was contrasted against the corresponding physical, economic and social indicators of the water footprint of the white chickpea crop produced in the same RDD, Cajeme, Sonora.

Based on Mekonnen and Hoekstra's (2011 *OP. CIT.*) methodology of recording and accounting of water consumption, the **definition** of the concept of *productivity* is used as a ratio of change, i.e. a quotient, in which the numerator is the quantity of product "Q" of physical or economic or social type, while the denominator is the quantity of water, thus obtaining the following mathematical model of water productivity:

$$\text{Productivity} = \frac{\underline{\text{quantity of product}}}{\text{WATER UNIT}}$$

Whereas water use efficiency was defined as a quotient in which the numerator is the amount of water used and the denominator is the unit of product (physical, economic, social...); therefore, the equation that would give rise to it would be the following:

$$\text{Efficiency} = \frac{\text{quantity of water}}{\text{product unit}}$$

It was **defined,** a priori, that a **permanent job** is equal to the number of working days that a human being works during a year in conditions of average social productivity, for which it was also assumed a *priori* that the human being works six days a week for 48 weeks a year, that is to say, that a permanent job is equivalent to the work that a human being performs in a year, equivalent to 288 working days, or in other words, a human being works in conditions of average social productivity, a total of 2304 hours a year (= 8 hours x 6 days a week x 48 weeks a year).

4.3 Methodology and mathematical models used

We used the methodology of the economic science, that is to say, the economic-mathematical logic, crystallised in models or mathematical equations, which describe each of the forms of productivity and/or efficiency (see this chapter, the part where we talk about the mathematical equations used, which go from Y_1 to Y_{14}) of water use, based on the methodology used by the Econom^a according to Ferguson and Gould[41] , which according to their table:

Figure 6: Methodology in the biological-agronomic sciences and in economic science
Source: Ferguson and Gould (1987)

Ferguson and Gould (1987 *Op. Cit.*), point out that economic science has its own scientific method, which is not based on experimentation (unlike biological and agronomical sciences for example), but on mathematical logic, i.e. on the abstraction of all irrelevant characteristics, and taking up only the essential characteristics of an economic phenomenon. It follows that economic science, unlike the biological sciences, has its own method, which is not based on experimentation, given the social nature of economics, but is based on logic.

Likewise, when analysing a single agricultural year and comparing two different crops, Descriptive Economics, one of the three scientific branches of Economic Science,[42] applied the *static-comparative* methodological approach - static because it was a single year and comparative because it was comparing several crops within that year (Astori, 1984)[43] .

The volume of irrigated water per ha is given by the model:

$$V=10000LR$$

Six independent **variables** were evaluated for the determination of the Imdrica footprint:

1) The amount of kg produced per cubic metre of water (Kg m^{-3}), generated

[41] **Fergusson CH. E. and J. P. Gould. 1987.** Theona microeconomica. FCE. Mexico, DF.pag. 11

[42] The other two branches are PoHtica Economia and PoHtica Economica. See Astori D, 1984. Critical approach to social accounting models. 5ª edicion. Siglo veintiuno editores. Mexico.

[43] **Astori D. 1984.** A critical approach to social accounting models. 5th edition. Siglo veintiuno editores. Mexico.

by the model:

$$kg\ m^{-3} = 10^{-1} *(RF/LR)$$

Where RF = physical yield per hectare (in ton ha-1), LR = Irrigation Rate Sheet

2) The amount of irrigation water required to produce one kg of physical product, generated by the model:

$$Litres\ kg^{-1} = 10^4 *(LR/RF)$$

3) The amount of profit in US dollars produced per cubic hectometre of irrigated water generated by the model:

$$USD\$\ of\ profit\ hm^{-3} = 10^2 * g* LR^{-1}$$

Where: g = profit per ha. g = RM - c = RF (p) - c, RM = monetary yield or revenue per hectare, "p" is the PMR and "c" is the cost per hectare.

4) The amount of irrigated water (in m^3) needed to produce USD\$1 profit, generated by the model:

$$m^3\ g^{-1} = 10^4\ LR\ g^{-1}$$

Where: 10^4 is the 10,000 m^2 of a ha.

5) The number of permanent jobs "E" associated with the use of one cubic hectometre (hm^3) of water used in irrigation, generated by the model:

$$Jobs\ hm^{-3} = (25/72)*(J/LR)$$

Where: J = number of working days per hectare. 1 working day = 8 working hours per day, it is assumed a priori that a person works 6 working days per week for 48 weeks per year, equivalent to 288 working days per $year^{-1}$; (25/72) is the simplification of the rule of three, in which we have, on the top left, "J" and on its right the amount of irrigated cubic metres per hectare (equal to the product of 10,000 m by the irrigation sheet "LR" in metres) and below, on the left side, "E" and on its right side 1,000,000 m^3 .

6) The private appropriation of profits in relation to the price paid for water, generated by the model:

$$g\ /\ price\ of\ water = g\ m^{-3}\ /cost\ per\ m^3\ of\ water$$

Where: the "water price" is the division of the item "irrigation" within the components or items in the cost structure "c" of production per hectare.

Other variables relating to capital and labour productivity were also analysed, which were as follows:

On the productivity of capital:

7) Benefit-Cost Ratio (BCR), estimated by the model:

RB/C = RM / c

8) Profit rate, estimated by the model:

Rate of return = (RM - c)/ c

9) Number of "E" jobs generated per million dollars, estimated by the model:

E/million USD = $(10^6 *(J/(288)))/c$

Where: 10^6 is one million US dollars.

10) Break-even point "EP", estimated by the model:

PE= c/RM =Cost ha^{-1} /RM ha^{-1}

On the productivity of capital:

11) Number of labour hours "h" invested per hectare, "h ha-^{1} ", estimated by the model:

$$h\ ha^{-1} = J*8$$

12) Number of working hours invested per tonne, "h ton^{-1} ", estimated by the model:

$$h\ ton^{-1} = J*8/RF$$

13) Kilograms produced per hour of work, "kg h^{-1} ", estimated by the model:

$$Kg\ h^{-1} = 10^3\ RF/(J*8)$$

Where: 10^3 is to convert RF, in ton ha^{-1} , to kg ha^{-1} .

14) Profit generated per hour of work, estimated by the model:

$$USD\ of\ gain\ h^{-1} = g/(J*8)$$

15) Profit generated per worker, generated by the model:

$$USD\ profit\ worker^{-1} = 288*g*J^{-1}$$

V. RESULTS AND DISCUSSION

15.1 Production, prices, income, cost and profitability of pump-irrigated wheat and chickpea grain crops in Cajeme, Sonora, OI 2014-15 agricultural cycle.

According to SIAP (2016) figures, a total of 20, 710, 981.57 hectares were harvested nationally in the 2014-15 OI cycle, of which 115, 550.88 were grain chickpea, representing 0.6% of the national harvested area, while grain wheat was harvested a total of 634,240.99 hectares, representing 3.1% of the total area. Likewise, the Gross Value of Production (GVP) generated by the entire agricultural sector was $395,508.06 million current pesos, of which the white chickpea crop accounted for 0.7% ($2, 622.67 million pesos), while wheat accounted for 3.0% ($11, 923.68 million pesos) of the national GVP.

Disaggregating these figures at the state level, it is observed that in the whole state of Sonora a total of 573, 765.63 hectares were harvested, of which 24, 657 hectares harvested were of white chickpeas (4.3%), and 304, 547.50 hectares harvested of wheat grain, which represented 53.1% of the harvested area at the state level. Likewise, agriculture at the state level generated a total of $27, 125.28 million pesos, of which 3.2% ($862.54 million pesos) were generated by the cultivation of chickpeas and 27.2% ($7, 384.39 million pesos) of grain wheat.

Likewise, according to SIAP figures (2014 *Op. Cit.*), at the regional level in Cajeme, Sonora, a total of 201, 179.04 hectares were harvested that year, of which 10,110 were grain chickpeas, representing 41.0% of the state area harvested of chickpeas, while grain wheat was harvested a total of 191, 068.00 hectares, representing 62.7% of the harvested area of grain wheat at the state level. Likewise, the Gross Value of Production (GVP) generated by the entire agricultural sector was $4, 839.72 million current pesos, of which the white chickpea crop accounted for 40.3% ($347.36 million pesos) of the value generated at the state level, while wheat accounted for 65.5% ($4, 839.72

million pesos) of the GVP generated at the state level (Table, 1).

Table 1: Relative importance of chickpea and wheat production in Cajeme.

Level of aggregation	Harvested area (ha)	Annual Physical Production (ton)	VBP (Millions of current pesos)
National			
All crops	20,710,981.57		$ 395,508.06
Chickpea grain	115,550.88	209,941.46	$ 2,622.67
Wheat grain	634,240.99	3,357,306.00	$ 11,923.68
% of chickpeas in relation to national	0.6%		0.7%
% wheat in relation to national wheat	3.1%		3.0%
State:			
Sonora	573,765.63		$ 27,125.28
Chickpea grain	24,657.00	65,670.88	$ 862.54
Wheat grain	304,547.50	2,089,841.40	$ 7,384.39
% chickpea in relation to the state	4.3%		3.2%
% wheat grain in relation to state	53.1%		27.2%
REGIONAL:CAJEME			
WHEAT	191,068.00	1,357,041.80	$ 4,839.72
GARBANZO	10,110.00	26,125.45	$ 347.36
% chickpea in relation to the state	41.0%	39.8%	40.3%
% wheat grain in relation to state	62.7%	64.9%	65.5%
The two crops	201,179.04	1,383,168.30	$ 5,188.13
% compared to national	1.0%		1.3%

Source: Own elaboration, based on figures from SIAP, 2014.

As mentioned in the previous paragraph, a total of 201,178 hectares were harvested in Cajeme, of which 10,110 were of white chickpeas (5%), while the remaining 95% were harvested of wheat grain (191,068 hectares), which produced a total of 26,125.50 tons of white chickpeas and 1,357,042 tons of wheat grain. The Gross Value of Production generated by these two crops for the Cajeme, Sonora region was US$310.3 million, US$20.78 million generated by the chickpea crop and US$289.54 million generated by the grain wheat crop.

Regarding the physical yield, Table 2 shows that 6.9 ton ha^{-1} of grain (wheat and chickpea) were produced, however, a breakdown of the figures shows that wheat had a yield of 7.10 ton ha^{-1} , while white chickpea had average yields of 2.58 ton ha^{-1} for that agricultural cycle, which indicates that the wheat crop had better yields that year (Table, 2).

On the other hand, Table 2 shows the price per ton of each of the two crops, finding that wheat had a price of US$213.4 per ton, while chickpea grain reached US$795.4 per ton, which indicates that the chickpea crop had a price 272% higher than that of wheat in Cajeme, Sonora. This price difference per ton caused the income per hectare for chickpea to reach US$2,055 dollars per hectare, while the wheat crop obtained US$1,515 dollars per hectare, during that agricultural cycle. On the other hand, the cost per hectare indicates that while in white chickpea was US$1,443 dollars per hectare, in wheat it was US$1,580 dollars per hectare, which indicates that the cost per hectare was US$1,443 dollars per hectare, while in wheat it was US$1,580 dollars per hectare.

hectare, which indicates that the cost per hectare in the chickpea crop was 8.7% lower than the cost in the wheat crop.

Table 2: Area, annual physical production, Gross Value of Production (GVP), Benefit-Cost Ratio (B/C), labour hours per tonne, employment generated and water used in irrigation in chickpea and wheat grain crops produced in Cajeme, Sonora. OI 2014-2015.

Macroeconomic variable	Wheat, Cajeme BFM. Mp	Garbanzo, Cajeme BFM. Mp	Total
a) Area harvested (ha)	191,068.0	10,110.0	201,178.0
b) Annual production (tonnes)	1,357,042	26,125.5	1,383,167.3
c) GVP (US$ million)	$ 289.54	$ 20.78	310.3
d) Physical yield "RF" (Ton/ha) = b/a	7.10	2.58	6.9
e) Price (US$)/tonne = 1000000*c/b	$ 213.4	$ 795.4	$ 224.4
f) Income (US$)/ha =d*e	$ 1,515	$ 2,055	1,542.5
g) Cost (Us$) /ha	$ 1,580	$ 1,443	1,573.3
h) Cost (US$) /ha without ground rent	$ 1,161	$ 1,024	1,154.6
i) Profit (US$) /ha= f-g	$ -65	$ 613	- 30.8
j) Benefit/cost ratio = f / g	0.96	1.42	0.98
(k) # of "J" days/ha	1.20	1.40	1.2
l) kg / day = 1000 d/ k	5,919	1,846	5,681.9
m) Cost (US$)/tonne = g/d	$ 222	$ 558	228.8
n) Profit (US$) /day = i /k	$ -54	$ 438	- 25.5
o) Net irrigated area (LR) in m	0.75	0.419	0.7
p) Volume of water used / ha (m)^3	7,500.0	4,190.0	7,333.7
q) Volume of water used over the entire harvested area (hitf)	1,433.01	42.36	1,475.4
r) Total monetary gain (Millions of US$)	$ -12.39	$ 6.20	- 6.2
s) Total daily wages per year	229,282	14,154	243,435.6
t) Number of permanent jobs/year	796	49	845.3
u) Capital invested in production (US$ million)	$ 301.94	$ 14.58	316.5

Source: 1) The area, production and nominal PPV (in thousands of pesos) and price per ton (nominal pesos / ton), come from SIAP, subsequently all monetary figures were expressed in US$ of 21 September 2015 at 1404 hours reported by the Bank of Mexico for the interbank FIX dollar with a parity of $16.175 per US dollar. 2) The cost per hectare, number of labourers and volume of irrigated water per hectare in commercial planting are from FIRA's Cost of Production 2014).

These differences in production costs per hectare and income per hectare are expressed in the different Benefit-Cost Ratios (B/C) that the crops had, while chickpea had a B/C ratio equal to 1.42, wheat had an indicator equal to 0.96, which indicates that while the producer dedicated to produce chickpea of each invested dollar, obtained that dollar and 0.42 additional dollars, that is to say; the production of chickpea was profitable, while the producer dedicated to the production of wheat, only recovered 0.96 dollars of each invested peso, which implies that during that agricultural cycle the production was not profitable. At this point it is important to mention that these profitability indicators are considered for producers who are renting the land at a given moment, however, if the producer were the owner of the land, the R B/C indicator for the wheat crop would be 1.30 and 2.01 in the case of chickpea.

Another of the variables indicated in Table 2 is the number of day labourers per hectare, detailing that while in the chickpea crop 1.40 day labourers per hectare were employed in the wheat crop, 1.20 day labourers per hectare were employed.20 daily wages per hectare, so that according to the calculations made it is estimated that during that agricultural cycle for the entire harvested area a total of 14, 154 daily wages were employed in the chickpea crop (which represents 49 permanent jobs per year), while the wheat crop required a total of 229, 282 daily wages, which represented 796 permanent jobs. Seen from another angle, the wheat and chickpea grain area required 1,475.5 hectometres of water (1,433.01 hm^3 for wheat cultivation and 42.36 hm^3 to grow grain chickpea), which would imply an expenditure of 7,500 m^3 per hectare in wheat and 4, 190 m^3 in chickpea, considering a net irrigation sheet, shown in Table 3 (based on FIRA production cost figures per ha), of 0.419 cm in grain chickpea and 0.75 cm in grain wheat (see Annexes 1 and 2), which when multiplied by the harvested area yielded the volume of water mentioned.

According to Table 3, it shows the absolute and relative cost structure in the cultivation of wheat grain and white chickpea in Cajeme, Sonora. From this source, it can be seen that the cost per hectare for chickpea was US$1,443 dollars per hectare, that is, $24,113 pesos per hectare. A breakdown of these figures shows that soil rent represented 29% ($7,000 pesos) of the total cost, and plant health represented 13.4% ($3,227 pesos). On the other hand, in the wheat crop, soil rent represented 26.5% ($7,000 pesos) of the total cost, followed by fertilisation with 19.2% ($5,078 pesos) of the total cost and miscellaneous items 13.1% ($3,469 pesos). This same analysis shows that the cost of irrigation represented in relative terms 3.9% ($940 pesos ha^{-1}) in the chickpea crop and 6% ($1,575 pesos ha^{-1}) in wheat, which implies that the cost per cubic metre was US$ 0.012 in wheat and US$ 0.011 in chickpea, this indicator is of utmost importance as compared to other agricultural areas, the cost of water is € 0.21 m$^{"3}$ (which is equivalent to $ 3.94 m^{-3}) (Salvador, *et al.*, 2011)[44] .

The cost of water is a particularly important index especially in arid and semi-arid regions where crop area tends to expand. This index therefore indicates the irrigation strategies to follow and the crops that would be competitive under certain circumstances. These values show that the price of water in the northern regions of Mexico is very low compared to other agricultural regions of the world, which contributes to an inefficient use of the resource, and that these prices do not reflect the real value of water, since according to Takele and Kallenbach, (2001)[45] , water prices are important for the improvement of water demand and conservation.

Table 3. Production costs per hectare in the cultivation of chickpea and wheat in Cajeme, Sonora. OI 2014-2015. B = pumping; G = gravity; F = fertilised; M = improved seed; mp = own machinery.

Concept	Absolute terms			Percentage structure		
	Wheat, Cajeme BFM. Mp	Garbanzo, Cajeme BFM. Mp	Average	Wheat, BFM. MP	Chickpea, BFM. Mp	Average
Land preparation	$ 2,351	$ 2,761	$ 2,372	8.9%	11.5%	9.0%
Sowing	$ 1,890	$ 2,914	$ 1,941	7.2%	12.1%	7.4%
Fertilisation	$ 5,078	$ 3,096	$ 4,978	19.2%	12.8%	18.9%
Cultural work	$ 190	$ 430	$ 202	0.7%	1.8%	0.8%
Irrigation	$ 1,575	$ 940	$ 1,543	6.0%	3.9%	5.9%
Phytosanitary	$ 2,517	$ 3,227	$ 2,553	9.5%	13.4%	9.7%
Harvesting, sorting and packing	$ 890	$ 890	$ 890	3.4%	3.7%	3.4%
Marketing	$ 780	$ 264	$ 754	3.0%	1.1%	2.9%
Various	$ 3,469	$ 1,789	$ 3,385	13.1%	7.4%	12.9%
Subtotal	$ 18,740	$ 16,311	$ 18,618	70.9%	67.6%	70.8%
Financial cost	$ 674	$ 802	$ 680	2.6%	3.3%	2.6%
Land rent per cycle ($/ha)	$ 7,000	$ 7,000	$ 7,000	26.5%	29.0%	26.6%
Total cost per ha (MX$)	$ 26,414	$ 24,113	$ 26,298	100.0%	100.0%	100.0%

[44] Salvador, R.; Martinez, C. A.; Cavero, J. and Playan, E. (2011). Seasonal on Farm Irrigation Performance in the Ebro Basin (Spain): Crops & Irrigation Systems. Agricultural Water Management. 98 (4): 577 - 587.
[45] Takele, E., & Kallenbach, R. (2001). Analysis of the Impact of Alfalfa Forage Production under Summer Water-Limiting Circumstances on Productivity, Agricultural and Growers Returns and Plant Stand. Journal of Agronomy and Crop Science, 187(1), 41-46.

		$$ 1,443	$ 1,573			
Total cost per ha (US$)	1,580.26					
Volume of water used per ha (Thousands of m)³	7.50	5.00	7.37			
Price of m³ (Mexican pesos)	$ 0.2100	$ 0.1880	$ 0.2104			
Price of m³ (US$)	$ 0.012	$ 0.011	$ 0.012			

Source: Own elaboration based on FIRA production costs per hectare in Annexes 1 and 2.

15.2 Water productivity indicators in wheat grain and white chickpea produced in Cajeme, Sonora.

The analysis of water productivity is shown in Table 4, which shows the productive, economic and social indicators. Water use efficiency is one of the most widely used indicators in a wide variety of crops in Spain (Garda *et al.,* 2013[39] ; Lorite *et al.,* 2012[46] [47] ; Romero *et al.,* 2006[48]), however in Mexico there is very little information and in some crops no information at all. In the present study the physical efficiency indicator of the wheat crop in Cajeme was 0.947 kg m^{-3} , finding a lower index in the white chickpea produced in the same region with 0.617 kg m^{-3} , which shows a lower efficiency of the chickpea crop to convert water into grain, since it used a total of 1,621 L kg^{-1} , compared to the wheat crop that used 1,056 L kg^{-1} (Table, 1).

However, the values of the physical productivity index of wheat grain are similar to those determined by Usman *et al.,* (2012)[49] who for Rechna Doab and Punjab, in Pakistan determined an index of 0.94 kg m^{-3} in wheat grain, while Shabbir *et al.,* (2012)[50] , determined an average of 0.43 kg m^{-3} , an indicator that is below that determined in Cajeme, Sonora. Other authors, such as Brauman *et al.* (2013)[51] [52] [53] reported rates of 0.9 kg m-3 for the United

[46] Garda, J. G., Lopez, F. C., Usai, D., & Visani, C. (2013). Economic Assessment and Socio-Economic Evaluation of Water Use Efficiency in Artichoke Cultivation. Open Journal of Accounting. 2(2): 45-52.

[47] Lorite, I. J., Garda-Vila, M., Carmona, M. A., Santos, C., & Soriano, M. A. (2012). Assessment of the irrigation advisory services' recommendations and farmers' irrigation management: a case study in southern Spain. Water resources management, 26(8), 2397-2419.

[48] Romero, P., Garcia, J., & Botia, P. (2006). Cost-benefit analysis of a regulated deficit-irrigated almond orchard under subsurface drip irrigation conditions in Southeastern Spain. Irrigation Science, 24(3), 175-184.

[49] Usman, M., Kazmi, I., Khaliq, T., Ahmad, A., Saleem, M. F., & Shabbir, A. (2012). Variability in water use, crop water productivity and profitability of rice and wheat in Rechna Doab, Punjab, Pakistan. J. Animal Plant Sci, 22(4), 998-1003.

[50] Shabbir, A., Arshad, M., Bakhsh, A., Usman, M., Shakoor, A., Ahmad, I., & Ahmad, A. (2012). Apparent and real water productivity for cotton-wheat zone of Punjab, Pakistan. Pak. J. Agri. Sci, 49(3), 357-363.

[51] Brauman, K. A., Siebert, S., & Foley, J. A. (2013). Improvements in crop water productivity increase water sustainability and food security-a global analysis. Environmental Research Letters, 8(2), 24-30.

[52] Aiken, R. M., O'Brien, D. M., Olson, B. L., & Murray, L. (2013). Replacing fallow with continuous cropping reduces crop water productivity of semiarid wheat. Agronomy Journal, 105(1), 199-207.

[53] Gonzalez, R. F.; Herrera, P. J.; Lopez, S. T.; Cid, L. G. 2014. Water productivity in some agricultural crops in Cuba. Revista

States and 1.3 kg m^{-3} in China, on the other hand Aiken *et al.,* (2013)[45], determined rates ranging from 0.28 - 0.62 kg m^{-3} . This indicates that in the region analysed, water productivity levels are comparable to those of other wheat-producing regions; however, improvements in irrigation water management must still be applied to increase water productivity in the cultivation of wheat grain. In the case of chickpea, Gonzalez *et al.,* (2014)[46] in Cuba, the physical productivity of chickpea ranged between 0.34-1.27 kg m-3, while in maize it varied between 2.03 - 16.43 kg m^{-3} , which indicates that in the northwestern regions of Mexico great efforts are still needed to increase the productivity of white chickpea.

The indicator of the variable profit per hectometre shows that in the wheat grain crop produced in Cajeme, Sonora, an average of 115.63 m^3 was required to generate US$1 dollar of loss (the indicator was - 115.63), while in the chickpea crop a total of 6.84 m^3 was used to generate one dollar of profit. Viewed another way, this implies that in the case of white chickpea a profit of US$146,276.84 dollars was generated for each cubic hectometre used in irrigation, while in the case of wheat US$8,648.39 dollars was lost for each cubic hectometre used in irrigating the crop (Table, 4). Despite the importance of such indicators, little information is available on the economic efficiency generated per cubic metre of irrigation.

There are some works developed in the Mediterranean for vegetables, fruit trees, cereals and oilseeds; in this sense, some authors determined in wheat that the gross profit was € 0.23 m^{-3} (equivalent to US$ 0.26), while in sunflower and grain corn it was € 0.53 m^{-3} (equivalent to US$ 0.59) (Garda *et al.,* 2013[47]; Romero *et al.,* 2006[48]), while Al-Qunaibet and Ghanem, (2014)[49] for the Riyadh region, Saudi Arabia US$1.51 m^{-3} for wheat grain.

This shows that the cultivation of wheat grain was economically unproductive in relation to the wheat produced in other regions of the world, given that it used a large amount of water in the region (1,433.01 hectometres of water),

Ciencias Tecnicas Agropecuarias, 23(4), 21-27.

[47] Garda, J. G., Lopez, F. C., Usai, D., & Visani, C. (2013). Economic Assessment and Socio-Economic Evaluation of Water Use Efficiency in Artichoke Cultivation. Open Journal of Accounting. 2(2): 45-52.

[48] Romero, P., Garcia, J., & Botia, P. (2006). Cost-benefit analysis of a regulated deficit-irrigated almond orchard under subsurface drip irrigation conditions in Southeastern Spain. Irrigation Science, 24(3), 175-184.

Al-Qunaibet, M. H., & Ghanem, A. M. (2014). Economic Efficiency in Wheat Production, Riyadh Region, Saudi Arabia. Life Science Journal, 11(12

and the Profit/Cost ratio of wheat production was negative at 0.96. On the other hand, it should be taken into account that the amount of water that is required to invest to generate $1 dollar, because according to the results, 115.63 m^3 of water to generate $1 dollar of gross profit, which in this case was lost, so it is deduced that even if the investment of water for the production of wheat is increased, the profitability of the crop would not be favouring the income generated, which implies an unproductive use of water.

Table 4: Indicators of soil, water, capital and labour productivity in the production of chickpea and wheat grain crops produced in Cajeme, Sonora.

Economic Variable	Expressed in:	Wheat, Cajeme BFM. Mp	Garbanzo, Cajeme BFM. Mp	Total and/or average for the 2 municipalities
Indicators of the Mercantile Footprint:				
Physical Footprint : Physical productivity of irrigation water	kg m-3	0.947	0.617	0.932
Physical Footprint : Physical efficiency of irrigation water	litres kg^{-1}	1,056	1,621	1,073
Economic Footprint: Economic Productivity of Irrigation Water	USD$ profit hm- 3	$ -8,648.39	$146,276.84	$ 4,176.99
Economic Footprint: Economic Efficiency of Irrigation Water	m^3 of water for USD$1 of profit	$-115.63	$ 6.84	$ 239.41
Social Footprint: Social productivity of irrigation water	Jobs hm- 3	0.56	1.16	0.57
rate of private appropriation of profits in relation to the price paid for water	Dimensionless	- 0.69	13.01	- 0.33
Social productivity of capital (jobs generated per USD 1 million)	Jobs / million US$	2.64	3.37	2.67
Break-even point "EP	Ton ha-^{1}	7.406	1.814	7.013
Physical Performance / PE	base 1	0.96	1.42	0.98
Labour productivity:				
Hours of work invested per ha	h ha-^{1}	9.6	11.2	9.7
Working hours invested per tonne	h ton-^{1}	1.4	4.3	1.4
Kilograms produced per working hour	kg h-^{1}	739.8	230.7	710.2
Earnings generated per hour of work	US$ h^{-1}	$ -6.76	$ 54.72	$-3.18

| Profit generated per worker | US$ worker^{-1} | $ -15,567 | $126,082 | $ 7,331 |

Source: Own elaboration, based on figures in Table 1 and 2 and the mathematical models used. Monetary figures in US$ FIX (interbank) on 21 September 2015 at 1404 hours, at the rate of $16.715 Mexican pesos per US dollar reported by Banco de Mexico.

The private appropriation of profits index shows the relation between the income generated and the price per cubic metre.

Thus, the index for wheat produced in Cajeme was -0.69, which indicates that for every dollar that the wheat grain producer paid per cubic metre of water, only $0.69 cents of that dollar was obtained in return. On the other hand, in the chickpea crop, the index was $13.01, which indicates that the crop was much more productive in economic terms than wheat grain, since for every dollar invested in chickpea irrigation, the producer obtained a return of US$12.01 dollars. In this sense, De Estefano and Llamas (2012)[54][55] , determined an indicator of € 0.24 m^{-3} average for cereals. Therefore, Garda (2015)[51] mentions that on a global scale, irrigated agriculture is recognised as the sector that demands the highest volume of water, so farmers have a great responsibility in the conservation of the resource and it is critical that they make efficient use of it. Despite the fact that the Cajeme wheat crop shows unfavourable figures, a simplistic recommendation *should not be* made, such as deciding to stop producing this crop since it shows indicators that indicate its inefficient economic use of water, since it is a crop of *strategic importance* for any nation, it is basic for human food, however, what *should be done* is to improve through agricultural management, through technical and genetic research, to generate wheat that makes better use of water.

Social efficiency indicators

In terms of social water efficiency, which is the number of jobs generated per hectometre of water, the indicator for wheat was 0.56 jobs hm^{-3} and 1.16 jobs hm^{-3} for chickpea. This indicator is low in relation to other crops such as vegetables and fruit trees that require a large amount of labour for activities that are not carried out in other fodder crops or cereals. In this sense, Garcia

---

[54]De Stefano, L., & Llamas, M. R. (Eds.) (2012). Water, agriculture and the environment in Spain: can we square the circle? CRC Press. 301p.
[55] Garda, M. J. 2015. Towards precision irrigation in strawberry cultivation in the Donana area. Doctoral thesis. University of Cordoba. International Doctoral School in Agri-Food eidA3. Rabanales Campus. March 2015.

et al., (2013)[52] determined a rate that ranged between 24 - 62 jobs hm^{-3} in the production of vegetables and fruit trees, while the production of greenhouse crops generate up to 190 jobs hm^{-3} , while Rtes *et al,* (2015)[53] determined an average for forage crops in the Comarca Lagunera of 0.048 jobs hm^{-3} ranging from 0.037 jobs hm^{-3} in alfalfa and 0.076 jobs hm^{-3} .

The hourly productivity, i.e. the amount of labour hours invested per ton of product (wheat grain), indicates that on average 1.4 h ton^{-1} was required, which indicates that to produce a ton of wheat 1.4 hours of labour are required, while in the chickpea crop the indicator was 4.3 h ton^{-1} , which indicates that the wheat crop is more productive, as less labour hours are required to produce a ton of grain. According to (Dorward, 2013)[56][57][58] , there are two other ways of expressing labour productivity, for structural indicators, it can be measured by the gross value of output generated in relation to the number of people employed, by the number of hours worked. Thus, it is determined that each worker dedicated to the production of wheat grain in Cajeme, Sonora added US$126,082 dollars to the value of that production chain in that agricultural cycle, while in the case of the worker dedicated to producing wheat, US$15,567 dollars per worker was added. These indices are closely related to the quantity of wheat and chickpea produced as well as to the market price. In this respect, there is a general discussion on agricultural productivity seen as labour productivity, as generally impHcit or expHcit indicators related to crop productivity are used.

In this sense, it was determined that the profit per hour of work invested for the production of wheat grain in Cajeme, Sonora was - US$6.76 h^{-1} , which indicates that this crop was unproductive in relation to chickpea grain, which generated US$54.72 h^{-1} , which indicates that in economic terms chickpea was the one that showed to be more productive in the use of water.

The same Table 4 shows that under the same crop and market conditions, the

---

[56] Garda, J. G., Lopez, F. C., Usai, D., & Visani, C. (2013). Economic Assessment and Socio-Economic Evaluation of Water Use Efficiency in Artichoke Cultivation. Open Journal of Accounting. 2(2): 45-52.

[57] Rfos, F. J. L., Torres, M. M., Castro, F. R., Torres, M. M. A., & Ruiz, T. J. (2015). Determination of the blue light footprint in forage crops of DR-017, Comarca Lagunera, Mexico. Rev. FCA UNCUYO, 47 (1), 93-107.

[58] Dorward, A. 2013. Agricultural labour productivity, food prices and sustainable development impacts and indicators. Food Policy 39 (1): 40-50. Available at:
https://scholar.google.es/scholar?hl=es&as sdt=0%2C5&q=Dorward%2C+A.+2013.+Agricultural+labour+productiv
ity%2C+food+prices+and+sustainable+development+impacts+and+indicators.+Food+Policy+39+%281%29%3A+40
-50&btnG=

minimum amount required to produce to have a viable operation (break-even point) is 7.40 ton ha^{-1} in the case of wheat and 1.81 ton ha^{-1} for white chickpeas produced in Cajeme. Taking into consideration the production obtained in each of the crops, it is observed that wheat grain was located below the break-even point, with 7.102 ton ha^{-1} , which places it as an economically unproductive crop, while the white chickpea crop was located above the break-even point with 2.584 ton ha^{-1} . The variable that assesses the credit vulnerability of the crop, from the perspective of how many times the physical yield per hectare covers the break-even point. Thus, Table 4 indicates that in the case of wheat, the physical yield per hectare (7.102 ton ha^{-1}), was able to cover 0.96 times the 7.406 ton ha^{-1} that it had as break-even point; that is, the crop in this region did not produce 0.304 ton ha^{-1} to recover only what was invested. While in the case of the white chickpea it was able to cover only 1.42 times its break-even point, which also indicates that the crop had a Profit/Cost Ratio equal to 1.42, which implies that in Cajeme, during that agricultural cycle, only the chickpea was economically productive.

VI. CONCLUSIONS AND RECOMMENDATIONS

VI.1 Conclusions

Based on the methodology used and the results obtained with it, the *first* hypothesis is rejected, since the ttdric footprint in *physical* terms in the wheat grain crop produced in Cajeme, Sonora was lower than that determined in the white chickpea crop, since wheat used 1,056 L kg^{-1} and produced 0.947 kg m^{-3} , while chickpea used 1,621 L kg^{-1} and produced 0.617 kg m^{-3} .

Based on the methodology used and the results obtained with it, the *second* hypothesis is accepted, since the water footprint in *economic* terms in the wheat grain crop produced in Cajeme, Sonora, was lower than the one determined in white chickpea, because in wheat 115.63 cubic meters were invested to generate US\$ dollar of loss, and US\$-8,648.39 dollars were generated per hectometre of water used in the irrigation of wheat, while in white chickpea 6.84 cubic meters were used to obtain a profit of US\$ 1 dollar, and US\$146, 276.84 dollars were obtained per hectometre of water used in the irrigation of white chickpea.

The third hypothesis is accepted, as chickpea generated 1.16 jobs per hectometre, while wheat generated 0.56 jobs per hectometre.

VI.2 Recommendations

It is recommended to carry out this type of study in other wheat and white chickpea producing areas such as Michoacan, Chihuahua and Guanajuato, in order to determine which region is efficient in the production of this basic grain so that these regions are in charge of producing such an important foodstuff, as well as in those regions where wheat should be substituted, given its inefficiency in the use of water, by other crops that make better use of water. It is also recommended that efforts be made in other areas such as the generation of germplasm as well as new technologies that can make water use more efficient in the arid and semi-arid zones of our country, taking into account that these areas produce a significant amount of food, have low rainfall and have a high incidence of climatic contingencies that affect agricultural production.

For the sake of autarky, i.e. food sovereignty, it is advisable that Mexico should not depend on wheat produced abroad to feed the people as well as to supply the bakery and pasta industry, so that the institutions in charge of research and production promotion should focus on producing and promoting new varieties of wheat that require less work hours per tonne, that demand less water per kg produced, and in which the capital invested is more socially profitable, should focus on producing and promoting the sowing of new varieties of wheat that require fewer working hours per tonne, that require less water per kg produced, and in which the capital invested is more socially profitable, that is to say that it generates more employment per amount of capital invested.

The physical water footprint, that is, the one that evaluates the amount of water per kg of product, is already a powerful instrument that allows selecting those crops with a lower unitary water demand, but it must always be kept in mind that *sustainability* is based on three precepts: The ecological, the economic and the social, which is why Mexican agriculture must stop taking the soil (tonnes per hectare) as the only reference for productivity and the physical, economic and social components of the water footprint (litres per kg, profit per m^3 and jobs generated per Hm^3 respectively) must be incorporated into all economic analysis of agriculture, livestock and forestry activities, since, for something to be sustainable, it must necessarily be sustainable in the long term, it must also be sustainable in the long term, since, for something to be sustainable, it must necessarily be sustainable in the long term, it must be sustainable in the long term, This poses the enormous task of elaborating national and regional water footprint matrices for use in generating agricultural, livestock and forestry patterns that are capable of saving water in the present for future use, while increasing the value of production and employment, and thus social welfare.

LITERATURE CITED

Aiken, R. M., O'Brien, D. M., Olson, B. L., & Murray, L. (2013). Replacing fallow with continuous cropping reduces crop water productivity of semiarid wheat. Agronomy Journal, 105(1), 199-207.

Al-Qunaibet, M. H., & Ghanem, A. M. (2014). Economic Efficiency in Wheat Production, Riyadh Region, Saudi Arabia. Life Science Journal, 11(12).

Amaya, M. J. 2015. Grain wheat (Triticum vulgare) from Cajeme, Sonora and its ^dric footprint. Thesis. Universidad Autonoma Chapingo- Department of Zootechnics. Texcoco Edo de Mexico. 55p.

Astori D. 1984. A critical approach to social accounting models. 5^a edicion. Siglo veintiuno editores. Mexico.

Brauman, K. A., Siebert, S., & Foley, J. A. (2013). Improvements in crop water productivity increase water sustainability and food security-a global analysis. Environmental Research Letters, 8(2), 24-30.

CONAGUA, 2007. National Water Commission. Determination de la Disponibilidad de Agua en el Acuifero 2605 Caborca, Estado de Sonora. CONAGUA, 31p.

De Stefano, L., & Llamas, M. R. (Eds.) (2012). Water, agriculture and the environment in Spain: can we square the circle? CRC Press. 301p.

D^az, J.P. 2013. Mexican wheat comes from southern Sonora. Note from the Economist of 14 August 2013. Available at: http://eleconomista.com.mx/columnas/agro-negocios/2013/08/14/trigo-mexicano-viene-sur-sonora / last accessed 15 October 2015.

Donnadieu P, M Roux-Michollet, V Chastagnier . 1999. Philosophical Magazine A, **1999^Taylor** & Francis. Available in: A quantitative study by **transmission electron** microscopy of nanoscale precipitates Dorward, A. 2013. Agricultural labour productivity, food prices and sustainable development impacts and indicators. Food Policy 39 (1): 40-50.Available at: https://scholar.google.es/scholar?hl=es&as sdt=0%2C5&q=Dorward %2C+A.+2013.+Agricultural+labour+productivity%2C+food+prices+and d+sustainable+development+impacts+and+indicators.+Food+Policy+3 9+%281%29%3A+40-50&btnG=

in Al-Mg-Si alloys

FAO: Unlocking water's potential for agriculture Chapter 3. Why water productivity matters for global water development. Sustainable Development Department. Rome, Italy. 2003.

Fergusson CH. E. and J. P. Gould. 1987. Theona microeconomica. FCE. Mexico, DF.pag. 11

Garda, J. G., Lopez, F. C., Usai, D., & Visani, C. (2013). Economic Assessment and Socio-Economic Evaluation of Water Use Efficiency in Artichoke Cultivation. Open Journal of Accounting. 2(2): 45-52.

Garda, M. J. 2015. Towards precision irrigation in strawberry cultivation in the Donana area. Doctoral thesis. University of Cordoba. International Doctoral School in Agri-Food eidA3. Rabanales Campus. March 2015.

Gonzalez, R. F.; Herrera, P. J.; Lopez, S. T.; Cid, L. G. 2014. Water productivity in some agricultural crops in Cuba. Revista Ciencias Tecnicas Agropecuarias, 23(4), 21-27.

Hoekstra, A. Y. 2006. The global dimension of water governance: Nine reasons for global arrangements in order to cope with local water.

problems. Value of Water Report Series No. 20. UNESCO-IHE, Delft, Netherlands. Available at:

http://www.waterfootprint.org/Report 20Global Water Governance.pd f/

Hoekstra, A. Y.; Chapagain, A. K.; Waterfootprints of nations: Water use by people as a function of their consumption pattern. Water Resources Management. 21(1):35-48.

Hoekstra, A. Y.; Chapagain, A. K.; Aldaya, M.; Mekonnen, M. M. 2011. The water footprint assessment manual setting the global standard. Earthscan, London, UK. 227p.

Hussain, I.; Turral, H.; Molden, D. and Ahmad, M. (2007). Measuring & Enhancing the Value of Agricultural Water in Irrigated River Basins. Irrigation Science. 25 (3): 263-282.

INIFAP. 2004. The cultivation of white chickpea in Sonora. National Institute of Forestry, Agriculture and Livestock Research. Northwest Regional Research Centre. Costa de Hermosillo Experimental Field. Libro Tecnico No 6. 290p.

INIFAP. 2005. Coast 2005. New white chickpea variety for the Hermosillo Coast. Instituto Nacional de Investigaciones Forestales, Agricolas y Pecuarias. Northwest Regional Research Center. Costa de Hermosillo Experimental Field. Technical Leaflet No 28. 24p.

Kijne, J. W; R. Barker; Molden, D (eds.) 2003. Water productivity in agriculture: Limits and Opportunities for Improvement. CABI Publication, Wallingford UK. 322p

Lorite, I. J., Gartia-Vila, M., Carmona, M. A., Santos, C., & Soriano, M. A. (2012). Assessment of the irrigation advisory services' recommendations and farmers' irrigation management: a case study in southern Spain. Water resources management, 26(8), 2397-2419.

Mekonnen, M. M.; and Hoekstra, A. Y. 2010. A global and high-resolution assessment of the Green, blue and grey water footprint of wheat. Hydrology and Earth System Sciences. 14: 1259-1276.

McCullough, E. B., & Matson, P. A. (2011). Evolution of the knowledge system for agricultural development in the Yaqui Valley, Sonora, Mexico. Proceedings of the National Academy of Sciences, 201011602. 1-6p.

Mekonnen M. M. and A. Y. Hoekstra. 2011. The green, blue and grey water footprint of crops and derived crop products. Hydrology and Earth Systems Sciences 15:1577-1600.

Monreal, R., M. Rangel, J. Castillo and M. Morales. 2002. Estudio de cuantificacion de la cuanficacion de la recarga del acuifero de la Costa de Hermosillo, municipio de Hermosillo, Sonora, Mexico. Hermosillo: University of Sonora.

Rfos F. JL.; Torres, M. MA.; Torres, M. M.; Castro, F. R. 2014a. Economic and social productivity of water in sultana and table grapes in DDR-037, Sonora. Report of the 59th Annual Meeting of the Central American Cooperative Programme for Crop and Animal Improvement (PCCMCA). Managua, Nicaragua from 28 April to 3 May 2014. 128pp.

Rfc "s Flores, J.L., Rios Arredondo B. E., Cantu Brito J. E., Rios Arredondo H. E.,, Armendariz Erives S., Chavez Rivero, J. A., Navarrete Molina C., Castro Franco R. 2018. Analysis of the physical, economic and social efficiency of water in asparagus (*Asparagus officinalis* L.*) and* table grapes (*Vitis virifera)* of DR-037 Altar-Pitiquito-Caborca, Sonora, Mexico 2014. Journal of the Faculty of Agricultural Sciences. National University of Cuyo. Volume 5. No. 1 year 2015. ISSN printed 0370-4661. ISSN on-line 1853-8665. Mendoza, Argentina. pp. 101-122.

Rios, J. L.; Torres, M. M.; Castro, R.; Torres, M. A. 2015. Determination of the blue water footprint in forage crops of DR-017 Comarca Lagunera, Mexico. Journal of the Faculty of Agricultural Sciences. Universidad Nacional de Cuyo. Volume 47. No. 1 year 2015. ISSN printed 0370-4661. ISSN on-line 1853-8665. Mendoza, Argentina. pp. 93-107.

Rtes, F. JL.; Torres, M. M.; Ruiz, T. J.; Castro, F. R. 2014. Productivity and irrigation water efficiency in cotton (Gosspium hirsuntum) from DR- 017 Comarca Lagunera and DDR-148 Cajeme, Sonora. Report of the X Congreso Nacional Sobre Recursos Bioticos De Zonas Aridas. Bermejillo, Durango 1 October 2014. 206-214pp.

Romero, P., Gartia, J., & Botfa, P. (2006). Cost-benefit analysis of a regulated deficit-irrigated almond orchard under subsurface drip irrigation conditions in Southeastern Spain. Irrigation Science, 24(3), 175-184.

Salinas-Zavala, C. A., Salvador E, L. C., & Fogel, I. 2006. History of winter wheat crop development in five irrigation districts in the Sonoran Desert, Mexico. Interscience. 31(4): 254-261.

Salvador, R.; Martmez, C. A.; Cavero, J. and Playan, E. (2011). Seasonal on Farm Irrigation Performance in the Ebro Basin (Spain): Crops & Irrigation Systems. Agricultural Water Management. 98 (4): 577 - 587.

Sanchez, M. A. (2008). Social sector organisations in the Yaqui Valley. Frontera Norte, 20(40): 135-167.

Schoups, G., Addams, C. L., Minjares, J. L., & Gorelick, S. M. (2006). Sustainable conjunctive water management in irrigated agriculture: Model formulation and application to the Yaqui Valley, Mexico.Water Resources Research, 42(10).

Shabbir, A., Arshad, M., Bakhsh, A., Usman, M., Shakoor, A., Ahmad, I., & Ahmad, A. (2012). Apparent and real water productivity for cottonwheat zone of Punjab, Pakistan. Pak. J. Agri. Sci, 49(3), 357-363.

SIAP (2014). Food Infograffa. Sonora. Agricultural Information System - SAGARPA. 1ª edition. 70pp. Available at: http://www.siap.gob.mx/pdfjs/web/viewer.php?file=Sonora.pdf / (Accessed 20 September 2015).

SIAP. (2015). Agricultural Information System. Closing of agricultural production   by state. Available in: http://www.siap.gob.mx/cierre-de-la-produccion-agricola-por-estado/

SIAP, 2016. Anuarios estad^sticos de la production agropecuaria. Agrifood and Fisheries Information Service. SAGARPA-SIAP. Available at: http://www.siap.gob.mx/ / (last accessed August 25, 2015).

SIAP. (2015). Agricultural Information System. Closing of agricultural production   by state. Available in: http://www.siap.gob.mx/cierre-de-la-produccion-agricola-por-estado/ (Accessed 30 September 2015).

Takele, E., & Kallenbach, R. (2001). Analysis of the Impact of Alfalfa Forage Production under Summer Water-Limiting Circumstances on Productivity, Agricultural and Growers Returns and Plant Stand. Journal of Agronomy and Crop Science, 187(1), 41-46.

Usman, M., Kazmi, I., Khaliq, T., Ahmad, A., Saleem, M. F., & Shabbir, A. (2012). Variability in water use, crop water productivity and profitability of rice and wheat in Rechna Doab, Punjab, Pakistan. J. Animal Plant Sci, 22(4), 998-1003.

Printed by Books on Demand GmbH, Norderstedt / Germany